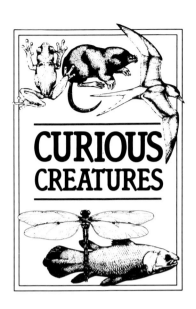

CURIOUS CREATURES

MISTAKEN IDENTITY

Written by
Joyce Pope

Illustrated by
Stella Stilwell and Helen Ward

STECK-VAUGHN
L I B R A R Y
A Division of Steck-Vaughn Company

Austin, Texas

Editor: Andy Charman
Designer: Mike Jolley
Picture Research: Jenny Faithful

Library of Congress Cataloging-in-Publication Data

Pope, Joyce.
Mistaken identity / written by Joyce Pope.
p. cm. – (Curious creatures)
Includes index.
Summary: Describes the many ways vulnerable animals protect themselves by making themselves difficult to see or mimicking better protected creatures.
ISBN 0-8114-3152-5
1. Camouflage (Biology) – Juvenile literature. 2. Mimicry (Biology) – Juvenile literature. [1. Camouflage (Biology)] I. Title. II. Series.
QL767.P67   1992   91-17136
591.57'2 – dc20   CIP AC

NOTE TO READER

There are some words in this book that are printed in **bold** type.
A brief explanation of these words is given in the glossary on p. 45.

All living things are given a Latin name when first classified by a scientist. Some of them also have a common name. For example, the common name of *Sturnus vulgaris* is common starling. In this book we use other Latin words, such as larva and pupa. We make these words plural by adding an "e," for example, one larva becomes many larvae (pronounced lar-vee).

Color separations by Positive Colour Ltd., Maldon, Essex, Great Britain
Printed and bound by L.E.G.O., Vicenza, Italy

1 2 3 4 5 6 7 8 9 0 LE 96 95 94 93 92

CONTENTS

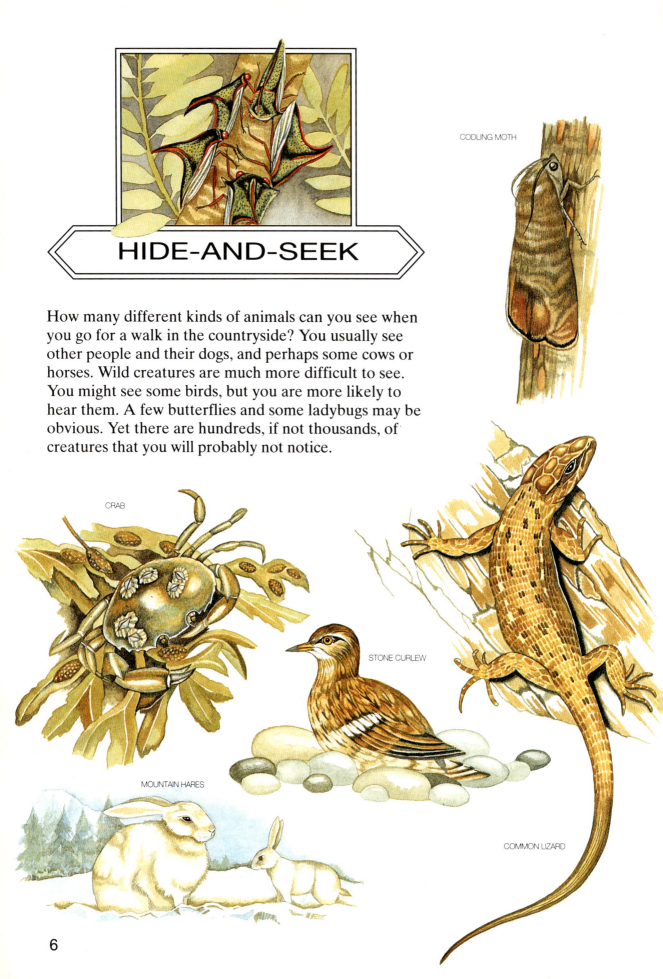

HIDE-AND-SEEK

CODLING MOTH

How many different kinds of animals can you see when you go for a walk in the countryside? You usually see other people and their dogs, and perhaps some cows or horses. Wild creatures are much more difficult to see. You might see some birds, but you are more likely to hear them. A few butterflies and some ladybugs may be obvious. Yet there are hundreds, if not thousands, of creatures that you will probably not notice.

CRAB

STONE CURLEW

COMMON LIZARD

MOUNTAIN HARES

6

Why is this? First, most of these animals are small. Some, like earthworms, live underground. But most creatures are invisible to you because they blend in with their background. We say that they are **camouflaged**. One other reason that we don't see wild creatures is that they keep still unless they are feeding or looking for a mate. We might spot them quickly if they were moving around.

Animals play a deadly serious game of hide-and-seek. This is because most of them live in a world where there is always the risk of being eaten. Both hunters and the hunted use tricks that help them stay alive. One of the best tricks is to be invisible. If they blend in with their background, their enemies (including you) will pass them by thinking that they are part of a tree, or perhaps a fallen leaf. If they are not seen, animals can escape being eaten by their **predators**. Predators may get a meal if they are not noticed by their **prey**. This book shows how animals disguise their identity with camouflage and other types of coloring.

▼ This is a tree frog from Brazil. If you were to explore its forest home you would probably not notice it, for it is colored and shaped like a dead leaf. Like dead leaves, it keeps quite still. You would probably be searching for something alive and active. Many small creatures depend for their safety on looking like something, such as a dead leaf, that most other animals are not interested in.

◀▼ All the animals shown here are easy to see. However, they disappear in the wild, for their shapes and colors match the patterns of the countryside where they live. Almost all animals, even very large ones, are camouflaged in some way. Camouflage hides them from their enemies.

ANGEL FISH

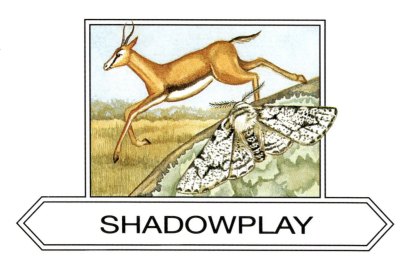

SHADOWPLAY

Although shadows are only patterns of dark and light, they make things stand out from their surroundings. A world without shadows would look flat. Many animals use a kind of camouflage that makes them look flat. Their coloring tricks the eye into not seeing the shadows made by their bodies.

COUNTERSHADING

Countershading is the name given to a color pattern that is found in almost all wild animals. Whatever they are, hunters or hunted, creatures of the land or sea, they

▼ These bonteboks were photographed in South Africa. They are good examples of animals that are countershaded. If you see bonteboks in a zoo, their color pattern seems very obvious. In the wild, their coloring makes the animals blend in with their surroundings as well as making them appear smaller.

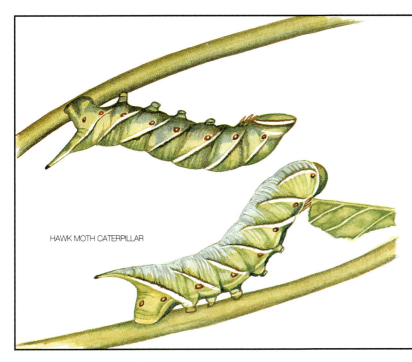

HAWK MOTH CATERPILLAR

Life Down Under
Many large caterpillars are so big and fat that they cannot sit on top of twigs as they feed. Instead they hang below. They live an upside-down life. Their underparts are much darker in color than their backs. This is called reverse countershading. Birds that hunt these caterpillars cannot see them if they are upside-down in their normal position. The birds find the caterpillars very quickly if the caterpillars are forced to sit on the upper side of the branches.

▶ A simple outline drawing of a cow makes the animal look flat. To make it seem more lifelike, you have to add shading. Then it appears to be solid and rounded. Countershading makes a solid and round animal look flat. This makes the animal hard to see and less likely to be caught and eaten.

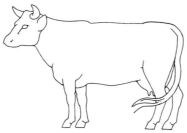

are paler on the undersides of their bodies than they are on their backs. Light falls on them from above, so their underparts are in shadow. The shadow darkens the pale underside, matching the upper body. This makes the animal look flat, rather than rounded. It seems simple, but it works, because a flat object merges into its background much more easily than a rounded one.

Many animals have other types of camouflage in addition to countershading. They may be blotched with color, or look like a part of their surroundings, but their backs almost always have more pattern and color than their bellies. Only a few animals are entirely white. They live in the Arctic, in caves, and in the deep ocean.

A moving animal is easy to see. Camouflage, of any kind, works best if the animal is still. Many creatures are able to keep perfectly still for long periods. As a result they remain unnoticed, even if they are out in the open. Often they let an enemy get very close, and then startle it as they rush off at the last moment.

USING SHADOWS

Wherever you are in the light, your shadow follows you. Even if you are very well camouflaged, your shadow can be a giveaway. Many animals hide their shadows. The simplest way of hiding a shadow is to lie as flat on the ground as possible. The closer an animal is to the ground, the smaller its shadow will be. Many birds do this. Baby **waders** and gulls, which are colored like bits of stone, flatten themselves when there is danger. As a result, a predator will not see a bird-shaped shadow, but something that looks like a stone. **Mammals**, such as small antelopes, use the same trick. They crouch down

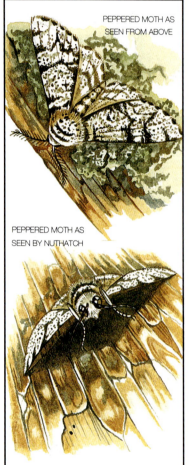

PEPPERED MOTH AS SEEN FROM ABOVE

PEPPERED MOTH AS SEEN BY NUTHATCH

Ghost in the Sand

Ghost crabs are colored so much like the sand in which they live that they are difficult to see. If a ghost crab is forced out of its burrow during the daytime, only its shadow is obvious as it scuttles across the beach. A hunting seabird follows this shadow. If a crab comes to a little hollow, it stops and folds its legs so that it is flat against the sand. Its shadow vanishes, and even the bird's sharp eyes cannot see the crab without its shadow. The crab then makes itself even safer by burrowing quickly under the sand.

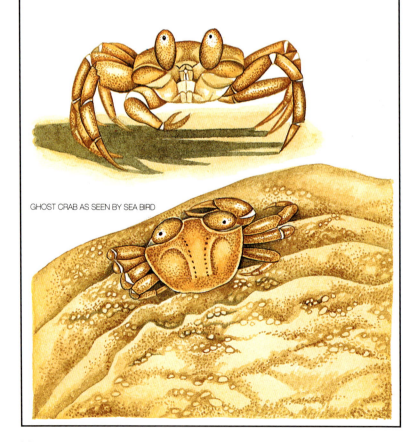

GHOST CRAB AS SEEN BY SEA BIRD

Catching Shadows

Peppered moths are colored like the bark of the trees on which they rest during the daytime. Their chief enemies are woodland birds, but the birds are usually deceived by the moths' camouflage which makes them difficult to see when viewed from above. Only one **species** of bird, the nuthatch, catches many peppered moths. This is because, unlike most small birds, the nuthatch walks up and down the trunk of the tree. The nuthatch does not see the color of the moth, but its shallow, V-shaped shadow. The camouflage that protects the moth against other hunters does not work in this case. Fortunately for the peppered moth and others colored like it, there are very few predators that hunt as the nuthatch does. Most moths escape.

and stretch their necks out along the ground. This makes them very hard to see, because the shapes and shadows of their horns and ears often match those of nearby plants and rocks.

Some caterpillars break up their shadows with fringes of hairs. A few kinds of lizards have long scales which also break up shadows. Many butterflies, when they rest, close their camouflaged wings and position themselves so that their shadow becomes no more than a thin dark line. Some butterflies lie almost flat on their sides, thus covering their shadows. Some other insects have false shadows in their coloring, making them look like plants or other parts of their **environment**.

▼ This leaf-tailed gekko from Madagascar catches insects at night. During the daytime it sits on the trunk or branch of a tree where it is nearly impossible to see for it is colored and patterned like the **lichens** covering its resting place. It is hunted by lemurs and hawks, but they rarely catch it. If it is seen by an enemy, it pretends to be fierce, by opening its mouth and flapping its tail, which is what it is doing in this picture.

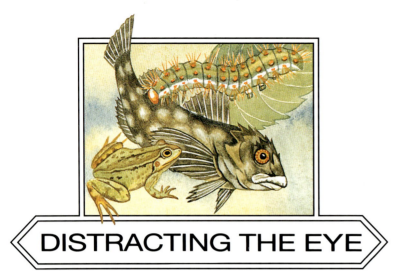

DISTRACTING THE EYE

NUMBAT

In the woods, the dappling of shadows moves as the wind shakes the leaves. In a grassy field there are constant changes in light and shadow as clouds pass over the sun. Even on an open stony plain, the pebbles and their shadows make the ground look like a jumble of different colors. Animals living in places like these have color patterns that make them blend in with their background. This makes them difficult to see.

MAMMALS

Mammals that move around at night are protected by the dark. They don't need to be strongly camouflaged.

▲▼ Here you can see some good examples of mammals with **disruptive coloration**. Large species are often blotched or have stripes that are vertical. Small creatures that run close to the ground usually have horizontal stripes. Some animals are disguised even more by special dark stripes around their eyes.

Breaking up the Outline

We recognize an animal mainly by its shape. A pattern of spots or stripes breaks up the outline of its shape and makes the animal difficult to see. Usually the pattern makes the animal seem like part of its background. At other times, it makes its shape meaningless to us. Experiments have shown that other animals see the same features as we do and are tricked in the same way.

CHIPMUNK

▲ Although tigers are now very rare in the wild, they live in many kinds of habitats. These include cold evergreen forests, hot forests, grasslands, mangrove swamps, and rocky places. They are difficult to see in all these places because their stripes break up their outline, making the animals merge with the colors and shapes of their background.

GROUND SQUIRREL

Animals that live in the sunlight must match their background. Countershading on its own does not make an animal completely safe. Blotches and stripes of color help to camouflage an animal even more. This is because the animal's background is also blotched and striped. Often these patterns break up the outline of the creature. When a hungry hawk sees a chipmunk, it sees stripes that could be pieces of stone or broken twigs. It is not only the hunted who must avoid being seen. Stripes help hunters, such as tigers, hide as well. A tiger's stripes make its outline merge with the shapes of tall grasses.

This breaking of the outline is called disruptive coloration. The spots on a leopard, the patchwork of a giraffe's skin, or the striped fur of a numbat all make the animals merge into their backgrounds. Young animals are more likely to get caught than adult animals because they are less wary and cannot run as fast. Young wild pigs and tapirs are camouflaged for protection, but they lose their disruptive coloration as they grow up.

WHEATEAR

BLUE TITMOUSE

WAXWING

AVOCET

SHELDRAKE

GOLDENEYE

▲ The bold black pupils in the eyes of many kinds of animals are easy to see. For this reason, birds' eyes are often disguised. They have a dark line running through them. You can see this line in birds and also in many kinds of mammals, reptiles, fish, and even insects.

Birds in flight are often difficult to see clearly because their outline is broken up by broad, pale colored stripes on their wings.

BIRDS

Birds that live in open country are often colored with patches and stripes of grays, browns, and white. This is called disruptive coloration. It breaks up their outline. Nightjars and snipes are colored in this way. They match their background very closely. They are almost impossible to see when they are resting.

A nesting bird does not leave its nest, because its eggs or chicks will die if they get cold. Most brightly colored birds rear their families in holes or in deep nests where they cannot be seen. Many birds, however, nest on the ground. They are usually so well camouflaged that they cannot be seen, even at very close range. If a predator comes close, the bird sits very still. At the last moment, it suddenly flies up. This usually startles the predator so much that it cannot find the eggs or chicks, which are also well camouflaged.

Sometimes only the female bird cares for the nest. The male is more brightly colored and more obvious. He may be seen by predators but at least his family is safe.

RINGED PLOVERS WITH CHICKS

REPTILES, FROGS, AND FISH

Most **reptiles** cannot run far when they are in danger. They can only survive by being difficult to see and by keeping still. Almost all are countershaded and many have disruptive coloration. Some lizards have disruptive patterns that disguise the shape of their heads.

Many snakes lie hidden on rocks or forest floors. They cannot hear airborne sounds and do not always slither away when a predator approaches. People are sometimes bitten when they step on a snake that they couldn't see.

Frogs are often brightly colored and have striped legs. These show up when the frog is moving. As soon as the frog is at rest, the disruptive leg pattern matches the pattern of its body. It's shape suddenly changes and predators are tricked.

Coral reef fish are often striped in bright colors. As they move their fins the patterns change and their shapes change with them.

▼ The Gaboon viper is a sluggish but very poisonous snake from western Africa. It may be responsible for killing as many as 1,000 people every year. A specimen that is 6 feet long has fangs about 11 inches long. The snake's disruptive coloration makes it almost invisible as it lies on the forest floor. The pattern on its back does not match the pattern on its sides and this makes it look less like a solid object. The viper's huge mouth is also disguised by a special area of disruptive coloration.

Black on White

The best kind of disruptive coloration breaks up the outline of an animal. Patches of black and white next to each other stand out in contrast to the less stark colors of the rest of an animal's body. These markings confuse predators. Many kinds of animals including fish, mammals, and birds have this type of camouflage. Chicks of ringed plovers have a white collar near black markings on their heads and bodies. As a result, from a distance, each chick looks like several small pebbles, and not like a bird at all.

INSECTS

The world is full of insects, but we do not notice most of them. This is because their bodies and wings are patterned to fit in with their backgrounds. Even brightly colored butterflies disappear from sight as they settle and close their wings. In most cases their undersides are colored with dull browns and grays. Preserved butterflies are shown with their wings fully spread out. This position is unnatural. The wings look very obvious. Living butterflies hold their wings partly folded so that lines and blotches of color break up their outline.

Some speckled moths, that move about at night, sit on the bark of trees by day. They do not match a particular background, but are almost invisible against tree trunks that are covered with lichens. In some places, the lichens have been killed by smoke and other types of **pollution**, and the trees have become black with soot. Over 100 species of moths living in these places have also become black. The black moths are hidden from the birds that hunt them. This color change is called **industrial melanism**. In places where pollution has been lessened, the lichens have returned and moths are once again light colored.

▼ Normal and black or melanistic forms of peppered moth.

▼ Grasshoppers are often noisy but they are difficult to see because their colors match their background. Sometimes the same species of grasshopper is differently colored in areas where there are different soils or rocks.

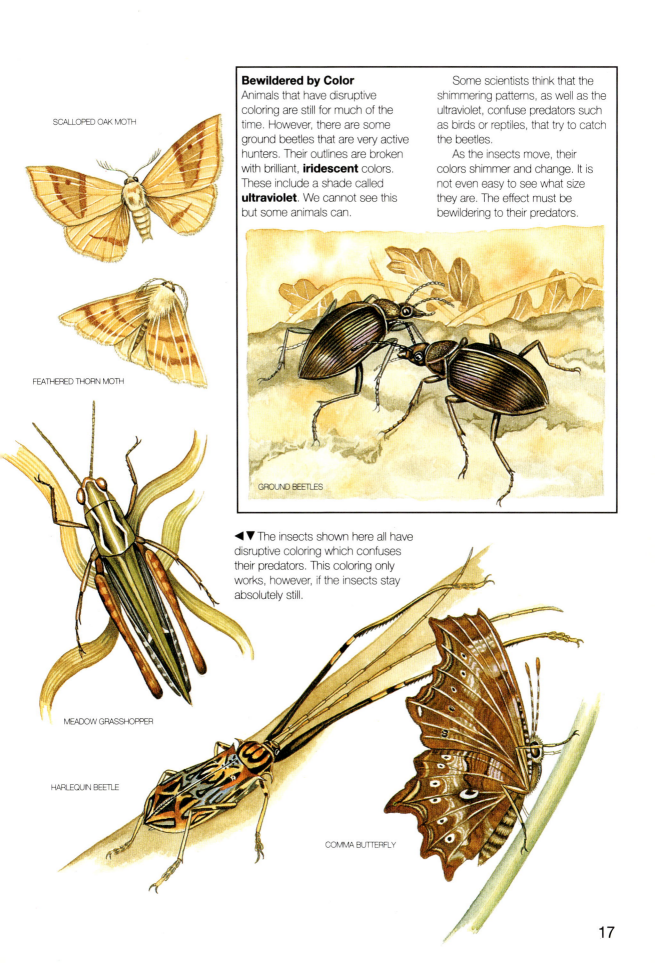

SCALLOPED OAK MOTH

FEATHERED THORN MOTH

MEADOW GRASSHOPPER

HARLEQUIN BEETLE

Bewildered by Color

Animals that have disruptive coloring are still for much of the time. However, there are some ground beetles that are very active hunters. Their outlines are broken with brilliant, **iridescent** colors. These include a shade called **ultraviolet**. We cannot see this but some animals can.

Some scientists think that the shimmering patterns, as well as the ultraviolet, confuse predators such as birds or reptiles, that try to catch the beetles.

As the insects move, their colors shimmer and change. It is not even easy to see what size they are. The effect must be bewildering to their predators.

GROUND BEETLES

◄▼ The insects shown here all have disruptive coloring which confuses their predators. This coloring only works, however, if the insects stay absolutely still.

COMMA BUTTERFLY

BROKEN OUTLINES

Some animals are difficult to see because their outlines are irregular. Their bodies are disguised with spines and knobs of tissue. They do not look like any normal living thing. This is called having a **disruptive shape** and it can be seen in many different kinds of creatures.

Coral reefs are the home of many kinds of animals. The spiky shapes of the corals make hiding places for them. Fish with spiny fins, creatures with spiny shells, and the waving tentacles of sea worms all merge into the background of corals. Stone and scorpion fish sit on the ocean floor. Their knobby surfaces make them look like small boulders. The animals that they prey on do not see them until it is too late and they are engulfed by the fish's huge mouth. People exploring the reef often do not notice them either. This can be dangerous, because their spines are connected to powerful poison glands.

On land, many insects break up the outline of their bodies. Spiny caterpillars are difficult to see because their shapes merge with the shapes of their background. Sometimes, broken outlines may make the insects look like crumpled dead leaves. Some grasshoppers have wings that look like partly eaten leaves.

▼ Animals that live in the edge of the sea are often brightly colored. Even so, they are often difficult to see because their shapes blend with rocks and fronds of seaweeds. Also, they often change shape. When they are feeding, they may stretch out tentacles. At other times, they are smaller and often look like pieces of rock or dead shells.

Spiders in Disguise

Some spiders change their normal shape by stretching their long legs along twigs. They are difficult to see because they look like parts of plants. Others use silk to break up their outline. Twig spiders from Australia sit on a small branch and look like the place where a twig has broken off. Others make zigzags of thick silk in their webs. Predators see the zigzags and miss the spiders.

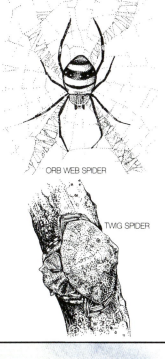

ORB WEB SPIDER

TWIG SPIDER

▲ This spiny caterpillar matches the colors of the twig and leaves that it is sitting on. The tufts of thick hairs on its back make it even more difficult to see. They break up its outline and may even look like small, growing leaves to the creatures that would eat the caterpillar if they found it.

19

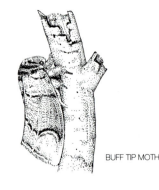

BUFF TIP MOTH

LOOK-ALIKES

Creatures with disruptive colors and shapes must match their background if they are to stay hidden from predators. An animal that is seen is likely to be eaten. Many creatures have, therefore, evolved outlines and color patterns that exactly copy some part of their environment.

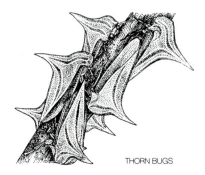

THORN BUGS

PRETEND PLANTS

How many leaves are not really leaves? In some trees there are insects that look just like leaves, twigs, or flowers. Insects sometimes imitate **blemishes** on the leaves. An animal has to appear flat to seem really leaflike. Some are shaded and colored in a way that makes them seem less round. Others are really flat. Some grasshoppers, fish, and lizards are narrow from side to side, like a book standing on a table. Others are flattened from above like a plate.

Indian leaf butterflies have hind wings that end in a long spike. When the butterflies sit on twigs the spikes look like leaf stalks. Their folded wings have a color pattern that looks like the **midrib** and veins of a leaf. Chameleons are the color of leaves. They sway as they creep up on their prey. In this way, they look and behave like part of a plant.

Many insects look like twigs. A few look like thorns and many look like bark that is covered with lichens. A result of these plant disguises is that unrelated creatures look very similar to each other. You may say that the disguises look good to us, but how do they appear to other hunters? Experiments have shown that birds are tricked by insects in the same way as we are.

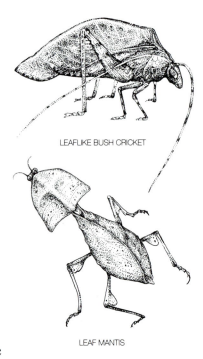

LEAFLIKE BUSH CRICKET

LEAF MANTIS

▲ Here you can see a collection of insects that look like plant parts. These insects survive because there are many more real leaves and twigs than there are creatures imitating them. A hungry bird or lizard learns very quickly the difference between a tidbit and part of a plant. If there were more leaflike insects than leaves, the disguise would not be so successful.

▶ Stick insects look like twigs. The leaf insect below lives in the forests of southeast Asia. The hind part of its body is very flat. Its wide wings are colored and marked like leaves. Sometimes a leaf insect swings from one leg, looking like a leaf about to fall.

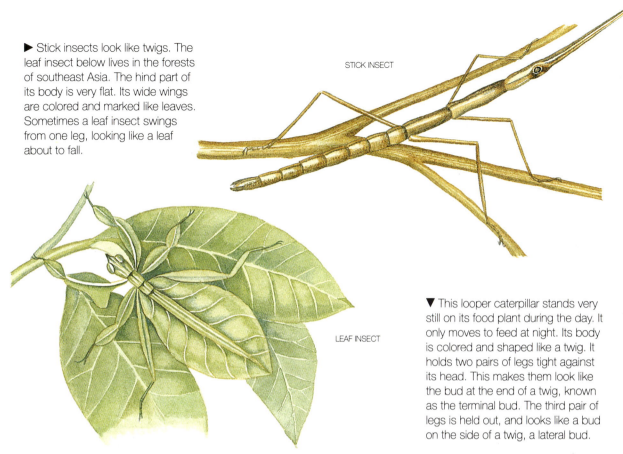

STICK INSECT

LEAF INSECT

▼ This looper caterpillar stands very still on its food plant during the day. It only moves to feed at night. Its body is colored and shaped like a twig. It holds two pairs of legs tight against its head. This makes them look like the bud at the end of a twig, known as the terminal bud. The third pair of legs is held out, and looks like a bud on the side of a twig, a lateral bud.

► The huge moth shown here is an atlas moth. It has a wingspan of almost 10 inches. Its home is the forests of southeast Asia. In spite of its size, it is difficult to see. When an atlas moth settles on the underside of a small branch and folds its wings it seems to change before our eyes. Its dull brown colors are the same as those of a dead leaf. It even seems to have holes, which make it look like a dead leaf that has been partially destroyed. Because its disguise is almost perfect, the atlas moth has little to fear from enemies while it rests.

Turning the Tables

The similarity between insects and flowers is not always because the insects mimic the plants. A number of orchids have flowers that look very much like insects. These plants have evolved to attract the insect to **pollinate** the flower. The flower often looks, and perhaps smells, like a female bee, wasp, or fly. The male insect attempts to mate with the flower and gets pollen on itself, which it then carries to the next flower. Other flowers may look like males, and are attacked by male bees trying to protect their territory. This also pollinates the flowers.

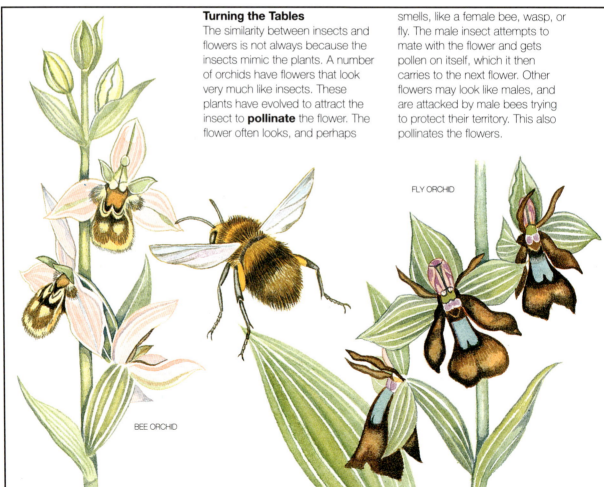

FLY ORCHID

BEE ORCHID

FISH AND TOADS

Some fish are disguised as dead leaves. The strangest of these is the leaf fish which lives in the backwaters of the Amazon River. It has a flattened body and a pointed, stalklike snout. It lies on its side or with its head downward, looking like a waterlogged leaf. In some species of leaf fish, the eye is disguised by dark lines.

The leaf fish feeds on other types of small fish. When it is hunting it seems to drift toward its prey. In fact, it is swimming very slowly, using very small movements of tiny, colorless fins. The prey is unaware of the danger until it is grabbed in the leaf fish's huge jaws.

Also living in the Amazon region are small leaf toads. Their heads are pointed and their bodies have a ridge of skin that makes them look like leaves. Below this, their sides are a dark color, like the shadows cast by leaves on the forest floor.

▲ The dragon fish that you can see here and on the cover of this book is a kind of Australian sea horse. It has flaps of skin that look like seaweeds. The dragon fish is slow-moving and helpless, but it is so well camouflaged that its enemies are unlikely to see it.

▼ Here you can see how the leaf fish and leaf toad are disguised. Leaf fish even have a dark line, like the midrib of a leaf, down their sides. They also have other markings that look like **mold** on a rotting leaf. The leaf toad sits on top of the fallen leaves of the forest floor and does not try to hide itself.

MIRROR ORCHID

LEAF FISH

LEAF TOAD

23

CHANGING COLORS

Two things make up an animal's color. One is chemicals, called **pigments**, which lie in or near the skin. The other is the type of surface of the animal's body covering. Some animals have a surface that breaks up and reflects the light that falls on it, making the animal shine in a rainbow of different colors.

COLOR MATCH

Many kinds of animals can change their colors. Animals that have shiny, reflective surfaces constantly change color as they move. This fools predators who

▶ This ermine has molted its brown coat which is replaced with a white one for winter. The change happens very quickly, because the white fur grows under the brown summer coat. This molts with the first heavy frost, so the ermine is well camouflaged for the winter.

▼ Ptarmigans are birds of northern moors and **tundras**. In the autumn, males and females molt and grow white feathers making them difficult to see in the winter snow. As spring comes, they molt again and grow speckled feathers which are a good summer camouflage. The females have completed their color change by May. They are nesting then and need to be invisible. The males do not take care of the young. Their molt is not complete until July, so their white feathers probably draw the attention of predators away from the females sitting on the nests.

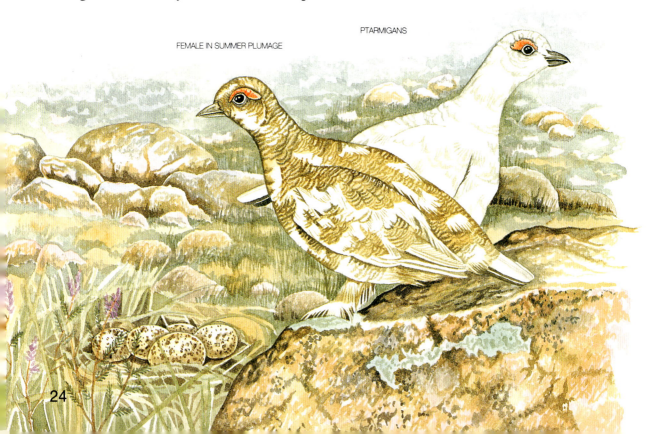

MALE IN WINTER PLUMAGE

PTARMIGANS

FEMALE IN SUMMER PLUMAGE

may, for instance, see something as purple one moment and dull brown the next.

Pigment colors also change. This usually happens slowly. Human skins, for example, tan in the sunshine. Mammals and birds may alter in color during a season as the tips of the hairs or feathers wear to show a different color from the parts nearer the skin. More often, color change is the result of molting hair or feathers. Some animals, such as foxes, replace their fur only once a year. Many others have a spring and an autumn molt. In the spring, birds molt their well-camouflaged winter plumage for bright feathers that will attract a mate.

Most Arctic mammals have an autumn molt which gives them a white coat. This camouflages them against a snowy background. In the summer months, their home changes color as plants sprout over the landscape. The animals also change color as they molt to a gray or brown color.

Animal Color

Normal white light is made up of a number of colors. Pigments are chemicals that absorb some colors and reflect all others. One of the most common pigments is **melanin**. This absorbs all light, so an animal with a lot of melanin appears black or dark brown. Carotenes, which are found in many plants, give red and yellow colors. There are many other, rarer pigments as well.

Interference colors are caused by light being broken up and partly absorbed by special **cells** on the surface of an animal. Sometimes these cells are ridged and sometimes filled with air or water.

QUICK-CHANGE ARTISTS

Most animals can only change color slowly, if at all. There are some, however, that can alter their appearance in less than a second. They do this by opening and closing bags of pigment, called **chromatophores**, which lie in the skin.

Each chromatophore has a branching shape, like a tree, and tiny muscles that open and close it. When the bag is open, the color is shown in every part of the bag. When it is closed, only a tiny spot of color shows.

We don't know exactly how an animal controls the opening or closing of the bag of color. In some cases, each chromatophore has its own nerve system directly controlled by the brain. The animal uses its eyes or spots in its skin to detect light and give the brain messages about the amount of light or the color of the background. The brain then sorts out what color changes are needed and directs the change.

Coral reef fish change the strength of their colors every day. They tend to be pale at night and more colorful by day, when there is plenty of light. Chameleons do the same thing. They tend to be paler in dim light and brighter in strong sunlight.

▼ This is a cuttlefish. Its chromatophores contain yellow, orange-red, and black pigments. The colors may be used on their own or together to make many different shades and patterns. It is almost impossible to see a cuttlefish in the sea, because it matches its background so well.

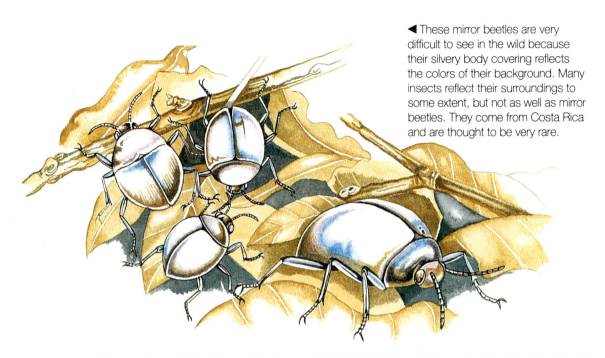

◄ These mirror beetles are very difficult to see in the wild because their silvery body covering reflects the colors of their background. Many insects reflect their surroundings to some extent, but not as well as mirror beetles. They come from Costa Rica and are thought to be very rare.

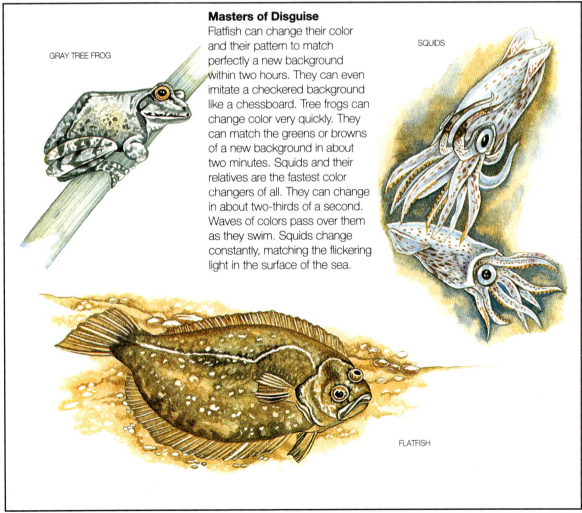

GRAY TREE FROG

SQUIDS

Masters of Disguise

Flatfish can change their color and their pattern to match perfectly a new background within two hours. They can even imitate a checkered background like a chessboard. Tree frogs can change color very quickly. They can match the greens or browns of a new background in about two minutes. Squids and their relatives are the fastest color changers of all. They can change in about two-thirds of a second. Waves of colors pass over them as they swim. Squids change constantly, matching the flickering light in the surface of the sea.

FLATFISH

THE COLOR OF FEAR

Have you ever blushed when you were angry or embarrassed? Blushing is the fastest color change in humans. When this happens, you get hot and your face turns red. **Hormones** in the blood make the tiny blood vessels just under the skin get bigger allowing them to carry more blood. The opposite happens when you are frightened. The tiny vessels close down and you go pale with fear. Similar things can be seen in many animals. A frightened chameleon often turns almost white; an angry one gets very dark. Frightened cuttlefish and octopuses often develop a black and white color pattern.

▶ The large photograph shows a chameleon in bright light, matching its background of green leaves. The chameleon in the small picture has seen a snake, which frightened it. It has puffed its body up and changed color, so that it contrasts harshly with its background. A hunting animal may be distracted by this sudden change in appearance.

▼ Cichlid fish fight in different ways. Some butt and bite each other. Some fight by displaying their colors. When one fish becomes exhausted it turns pale and leaves, tired but unhurt.

RED CICHLIDS FIGHTING

Many animals also change color during their breeding season. Many kinds of creatures, including birds, reptiles, **amphibians**, and fish get brighter when they are trying to attract a mate. These changes may last for weeks or even months. For instance, male frigate birds have a huge red pouch below their bills that they puff out during their mating displays. Male cichlid fish become more brightly colored as they fight for territory in which to breed.

The Octopus's Ink Blob
Octopuses have ink sacs that they use when they are frightened by an enemy, such as a moray eel. A frightened octopus changes its color many times. If the eel is able to follow it, the octopus turns dark and puffs ink into the water. This ink makes a black blob which is about the size of the octopus. The octopus then changes immediately to a pale color and jets off.

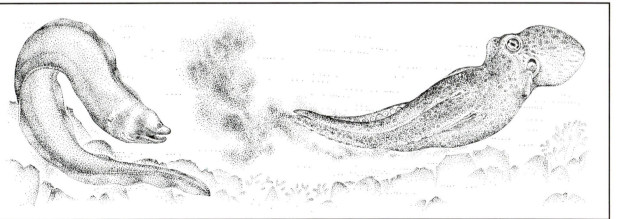

WARNING!

▼ Here are some insects that have warning colors. Wasps and bees can sting. Some bugs taste so bad that no animals will eat them. Ladybugs also taste bad. Ladybugs may bleed a little from special pores in their legs if they are attacked. This gives the predator a warning of an unpleasant meal if it persists.

Not all animals are camouflaged. Some are brilliantly colored. Some of these animals are attracting a mate; others use color as a warning. This is particularly true of animals that are black with red, yellow, or orange. These colors warn predators to leave them alone.

WARNING COLORS

In general animals that have warning colors are small creatures that are active in the daytime. From coral snakes to ladybugs, they seem to want you to notice them. Bees and wasps, protected by their stings, forage at length on flowers and fruit. Stinkbugs sit in the open regardless of danger.

You may wonder how the hunters of insects know that they should avoid these brightly colored species. The answer seems to be that once an animal is stung, or suffers the horrible taste of a ladybug or stinkbug, it remembers the experience and the bright colors that went with it. Experiments have shown this to be the case.

Animals with warning colors are often difficult to kill. A frog may try to eat a wasp and get stung, or chew a ladybug and discover the foul taste. The insect will probably escape without much harm and will have taught the attacker a lesson it will not forget.

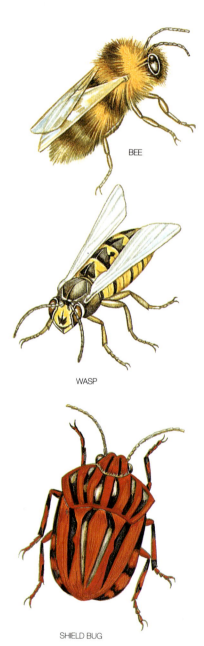

BEE

WASP

SHIELD BUG

TEN SPOT LADYBUG

SEVEN SPOT LADYBUG

Colors to Attract

Birds, reptiles, and fish are among the many kinds of animals that use bright colors to attract a mate. Often these colors are shown for a short time during the breeding season and are lost afterward. At other times it would be a disadvantage for the animal to be noticed. Often these colors are emphasized with a display, such as that used by a peacock showing off his tail fan, or a newt dancing to show off his bright underparts. Brightly colored males often flash their colors to warn rivals to stay away from their territory. These same colors mean something different to the females, which are attracted by them.

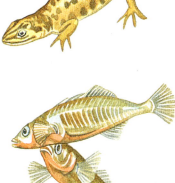

COMMON NEWT

STICKLEBACKS

▲ Bright colors are sometimes used for advertisement. This cleaner shrimp needs to be noticed by the fishes that are its clients. When a fish approaches, the shrimp climbs on to it and works its way over the scales and even into the fish's mouth and gills, removing parasites.

GREATER BIRD OF PARADISE

SHOCK TACTICS

A well-camouflaged animal is often helpless if it is discovered by a predator. Animals that are used to sitting still are unlikely to be able to run or fly fast once their disguise is seen through. Under these circumstances, some animals have a last card to play. They suddenly show a bright flash of color. Some lizards have a flap of bright skin; fire-bellied toads show their red underparts. This flash is sometimes enough to startle the predator for long enough to allow the animal to escape.

Insects and other animals that flash bright wings or legs are usually showing their colors to a predator that is not going to give up easily. They rely on the predator pausing for a moment. In that short time, the animal drops to the ground and scuttles away. The predator still seems to be looking for something brightly colored and often cannot rediscover the prey once it has hidden its bright colors again.

Sometimes the bright flashes are colored like eyes. The predator suddenly sees a big face, instead of a small meal. It may think that it is in danger of being eaten itself.

▼ Here are some animals that use shock tactics. The flying lizard spreads a brightly colored web of skin along its sides as it leaps for safety. Insects as varied as moths, cicadas, stick insects, and beetles may all use bright flashes of color to help them escape. Fire-bellied toads turn over to flash their red undersides.

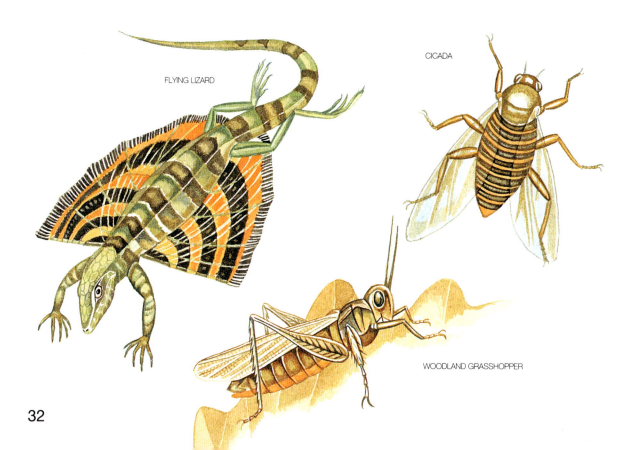

FLYING LIZARD

CICADA

WOODLAND GRASSHOPPER

▲ This lubber grasshopper is normally difficult to see as it sits on plants. If it is noticed by a small lizard or bird it takes off suddenly, displaying its brilliantly colored hind wings as it flies. The predator is likely to be confused by this and lose its meal as a result.

RED UNDERWING MOTH

FIRE-BELLIED TOAD

MIMICRY

The word "**mimicry**" means copying something closely. When we are talking about animals, we use it to describe the likeness between two animals that are not closely related. These animals are usually brightly colored and at least one of them is protected by a sting or bad taste. The colors are a warning to predators.

MODELS AND MIMICS

Mimicry has been discovered in many kinds of animals. Some plants are also mimics, sometimes of other plants, sometimes of animals. Mimicry is, however, most common and most dramatic among insects.

The animal that is protected by its unpleasant taste, smell, or sting is called the model. The animal that mimics the model must live in the same area to benefit from its likeness. The mimic must also be much rarer than the model. This is because predators must learn that the color pattern shared by the mimic and the model means that these animals are not good to eat.

There are many different kinds of mimicry. In one well-known case, a grasshopper that mimics a beetle is so much like the model that a museum collection had these two very different kinds of creatures mixed up. In many cases the likeness is not so great – just enough to trick a hunter.

Mimicry can be more than just looking the same. There is a beetle that mimics bumblebees by beating its wings at the same rate. This means that it makes the same buzzing sound. Other mimics smell like the models. Many behave in ways that remind predators of well-protected creatures.

DAY-FLYING MOTH (model)

SWALLOWTAIL BUTTERFLY (mimic)

MONARCH BUTTERFLY (model)

VICEROY BUTTERFLY (mimic)

▲ The illustrations on these two pages show that butterfly mimics are found all over the world. The day-flying moth from New Guinea tastes unpleasant. It is mimicked by a species of swallowtailed butterfly. Both are avoided by predators. The unpleasant-tasting monarch butterfly is mimicked by the viceroy butterfly.

▼ The two butterflies shown at the top of the picture below look alike, but are unrelated. They both come from Africa. The dark brown butterfly is a male. His mate is shown immediately below. She looks unlike him because she mimics another unrelated unpleasant-tasting species.

DANAUS FORMOSA (model)

PAPILIO REX (mimic)

MALE HYPOLIMNAS MISIPPUS

Henry Bates, Explorer

Henry Bates was an English naturalist who spent many years exploring the forests of Brazil. He made great collections of insects and other small animals. He also observed the creatures and wrote about their habits. He was the person who first noticed that some butterflies that looked alike were in fact unrelated. The type of mimicry in which one species copies another very closely is named after Bates. It is known as Batesian mimicry. Bates found many examples of mimicry in the tropics. Since his time, it has been discovered in far more kinds of animals than he observed.

FEMALE HYPOLIMNAS
MISIPPUS (mimic)

DANAUS CHRYSIPPUS (model)

BUTTERFLY MIMICS

Soon after Henry Bates first noticed mimicry in butterflies, another form of mimicry was discovered. A German explorer named Fritz Muller observed that well-protected animals often look more or less alike. Sometimes this is because they are closely related, like wasps and hornets; but this is not always the case.

Muller suggested that this similarity among a large number of species would help to protect them all. Predators would learn to avoid a whole group of species after very few attacks. If each pair of models and mimics were entirely different from all others, the predators would harm a greater number of individuals before learning to leave them all alone. The way this "defense club" behaves is known as Mullerian mimicry.

The butterflies that Fritz Muller first studied are called *Heliconiids*. They all taste bad and some kinds

▼ This Heliconiid butterfly from Brazil is one of a large group of species which all taste unpleasant. Many are colored orange and black, others are mainly black and white. Although easy to see and catch, they fly slowly, as if they know that nothing will eat them.

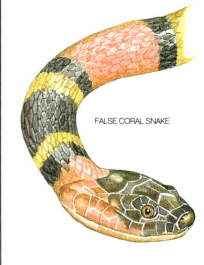

FALSE CORAL SNAKE

Sneaky Snakes
Poisonous coral snakes live in many warm parts of the world. They are brightly banded with red, yellow, and black. In some areas, snakes that look very similar to them live in the same places. They are known as false coral snakes, because they are not poisonous. This looks like straightforward mimicry, but that is unlikely.

are poisonous. A hunter such as a bird, lizard, or monkey has to eat only one of these butterflies to know that they taste horrible. Caterpillars of the poisonous Heliconiids feed on leaves that break down, when digested, to give the poisonous chemical hydrogen cyanide. The adults fly slowly and are easy to catch. They have tough bodies and can survive far more rough handling than most other butterflies. Groups of these butterflies roost together in the trees at night. The mass of bright color seems to scare off predators.

Some travelers in the Amazon area noticed pairs of Heliconiid wings on the forest floor. They discovered that the wings belonged to butterflies that fluttered to the ground after being damaged but not eaten by a predator. Foraging ants then seized the butterflies. The wings were snipped off by the ants' powerful mouthparts. The ants then took the bodies to feed their young. The unpleasant flavor did not seem to deter them.

False Eggs

Caterpillars of Heliconiid butterflies feed on the leaves of passion flower vines. Many species of these plants have developed defenses against the hungry caterpillars. Some species of the vine are extremely poisonous. One species has hairs on the leaves that trap and kill the caterpillars. Some even use a kind of mimicry against the egg-laying butterflies. On their leaves are small yellowish warts, about the same size and color as the butterfly's eggs. A female Heliconiid will not lay her eggs on these leaves. She looks for others where her offspring will not have to share their food.

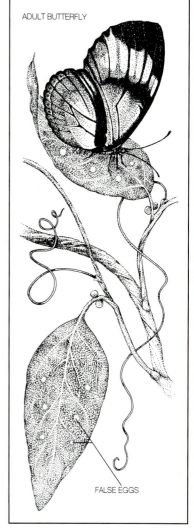

ADULT BUTTERFLY

FALSE EGGS

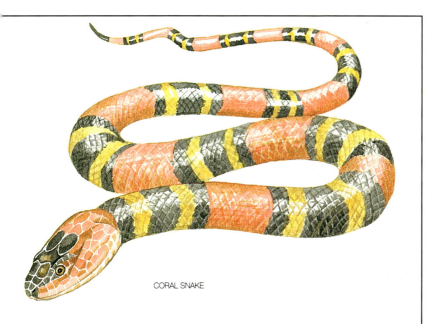

CORAL SNAKE

The poisonous coral snakes behave in different ways. The most **venomous** are placid animals, but their bite can kill even large animals. A predator that attacks one of them will not learn anything from the experience – it will be dead. Some of the less poisonous species are much more aggressive. They only kill and eat small rodents. Larger animals that attack them will, however, get a nasty surprise. The attacker will probably survive, but will leave red, yellow, and black banded snakes alone in the future. It is likely that both the deadly coral snakes and the non-poisonous, false coral snakes are both mimics of the mildly venomous, aggressive species.

WASP AND ANT MIMICS

Bees, wasps, and ants are protected by their stings. Many of these insects live in large family groups, which give them added safety. If the nest is attacked, all the insects will come to its defense. These insects are mimicked by many others. The yellow and black stripes that say "wasp" to us are worn by many harmless flies. Others, such as the drone fly, look like a honeybee. When these insects feed together on flowers in late summer, it is hard to tell which is the bee and which is the fly.

It is only the adult flies, however, that look like bees or wasps. In some cases, fly grubs are scavengers and grow in the safety of wasps' nests. It may be that the fly is able to enter the wasps' nests to lay her eggs, without being attacked, because she looks so similar to the real inhabitants of the nest.

Ants are mimicked by many creatures, including beetles and spiders. They use shading to mimic the effect of an ant's narrow-waisted shape. They often scurry like ants, tapping the ground with their **antennae** or, in the case of spiders, with their first pair of legs.

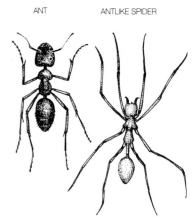

ANT   ANTLIKE SPIDER

▲ Ants are protected by stings and acid sprays. They are always part of a large group, all members of which help others if they are attacked. Because of this, many kinds of creatures mimic ants. Even some spiders do so. They hold up their front legs to look like antennae.

BUMBLEBEE

TRICHIUS

VOLUCELLA BOMBYLANS

DRONE FLY

Bee and Three Mimics
A bumblebee, with its striped, furry coat is easy to recognize. Most bumblebees are slow to sting, but they are well enough protected to be models for some mimics.

The hover fly *Volucella bombylans* mimics at least two forms of bumblebee. It has a furry, striped body. Details of its eyes and antennae show that it is not a bee.

Trichius is a beetle that imitates bumblebees very well. It sounds like a bee when it flies, because its wings beat at the same speed.

The drone fly mimics honeybees in its size, color, and the sound of its beating wings.

Pretend Stings

Some creatures that are quite unlike bees or wasps pretend to have stings. A hawk moth has a bundle of hairs which it sticks out at the rear end of its body. It may be spreading scent, but to an attacker, it looks like a stinger. A burying beetle imitates a cold bumblebee. It lies on its back and makes a buzzing noise. At the same time it moves as if it were about to sting, and gives off a stinking, frothy fluid.

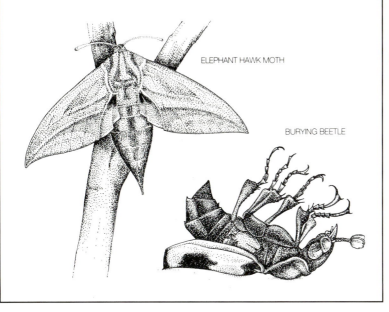

ELEPHANT HAWK MOTH

BURYING BEETLE

HOVER FLY

▲ The wasp in the photograph above is a model for a number of different mimics. They include the hover fly shown here. You might mistake it for a wasp. Many predators do, in spite of the fact that it flies differently.

LEAF BEETLE (model)

COCKROACH (mimic)

LADYBUG

LADYBUG MIMICS

There are about 2,200 different kinds of small beetles that are known as ladybugs. They are among the easiest beetles to recognize, because they are all brightly colored and round. The colors run from bright pink to red, orange, and yellow, usually with black or white spots. These colors are a warning to predators, because ladybugs have a bad taste.

Ladybugs do not hide from predators. They are active during the day and can be seen scurrying about plants looking for their prey. They eat aphids and other tiny creatures. Birds and other small predators avoid ladybugs. This is because ladybugs can give advance warning of the shock in store for anything that eats them. They are able to produce a tiny drop of their body fluids through special pores in their legs. One taste is enough to discourage the hungriest predator. The ladybug's smooth, rounded wing cases make the insect difficult to kill, so they walk off unharmed.

It is not surprising that ladybugs have many mimics. Some of these are leaf beetles, which are also foul tasting. Others probably taste good, but predators avoid them because they look like ladybugs.

◀▼ Here you can see some unpleasant tasting leaf beetles and ladybugs. They form a "defense club" as described on page 36, but both are mimicked by other insects.

In some parts of the world, cockroaches imitate both. The long antennae of the cockroaches give them away, but predators are unlikely to notice these.

LADYBUG

COCKROACH

COCKROACH

▼ Adult ladybugs **hibernate** through the winter. In some species, large numbers may gather together in suitable places. This is often in the open, because their colors defend them whether or not they are active. Many are killed during hibernation by cold or by fungus or virus diseases.

▼ Ladybugs almost always have contrasting spots on their brightly colored wing cases. The number of spots varies, some have as few as two, others as many as 28. They do not tell the age of the beetle. This species of ladybug lives in grass and in cornfields, where it feeds on aphids and also on pollen.

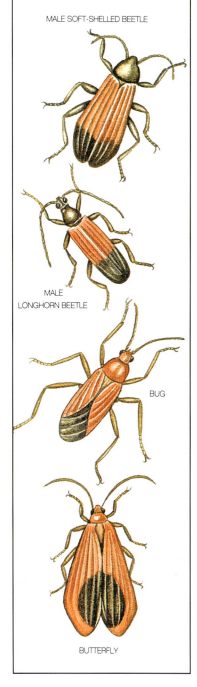

Join Our Club

These insects, which all look alike, are not closely related. They include two quite different types of beetles, a bug, and a butterfly. They all come from Borneo. It is probable that they all taste bad, so they form a defense club against predators including birds, small mammals, reptiles, and amphibians.

MALE SOFT-SHELLED BEETLE

MALE LONGHORN BEETLE

BUG

BUTTERFLY

VERTEBRATE MIMICS

Animals with backbones (known as vertebrates) rarely mimic each other. Some scientists believe that more mimics will be discovered as we learn more about animal behavior. One of the few cases of mammal mimicry is found in Borneo. There, small creatures called tree shrews are avoided by most flesh-eaters because they taste very unpleasant. They are mimicked very closely by squirrels, which can be eaten.

There are two species of flycatchers in Africa which are almost impossible to tell apart from some other birds. These other birds feed on ants, and taste of the acid that was part of the ants' bodies. This makes the birds taste so bad that no animals will eat them.

Several kinds of birds mimic snakes, some by hissing, others by snakelike movements of their heads and necks. There are some birds of prey that look like harmless species. The animals they catch do not suspect that they are being hunted until it is too late.

Barber fish remove parasites from bigger fish. They are mimicked by other species that feed on the fish wishing to be cleaned. The weever fish is protected by poison glands and is mimicked by the harmless sole.

TREE SHREW

SQUIRREL

▲ Here, you can see a tree shrew and a squirrel that come from the same area of southeast Asia. Both types of animals are active during the daytime, running and jumping about the trees. Apart from their long whiskers, which are probably only noticeable at close range, the edible squirrels could easily be mistaken for the unpleasant-tasting tree shrews.

The Crafty Cuckoo
Female cuckoos lay their eggs in the nests of other birds, which rear their young. In some countries such as Britain, they use many species of birds to bring up their families. In other regions, only one kind of bird looks after the cuckoos' eggs. Where this is the case, the cuckoos' eggs are colored like the eggs of the host.

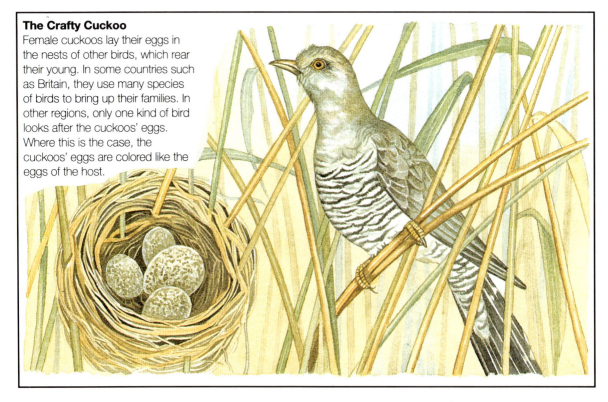

▲ This weever fish spends most of its time buried in the sand. It is invisible except for a dark patch of color around poisonous spines on its back. If a large fish or other predator approaches, the spines are raised to warn the predator of the danger of getting too close.

▼ Soles, like the one shown here, are found in shallow pools on the beach at low tide. They are protected by their camouflage, and by one small black fin by their head. Soles raise this fin if an enemy approaches. This is the same warning signal given by the weever fish, though the sole cannot back up its threat with poison.

HUMAN CAMOUFLAGE

Human beings usually do not fear enemies that may kill and eat them. We walk about openly, dressed in brightly colored clothes, making a lot of noise. Sometimes we need to use camouflage. If we want to watch how other animals behave, we may have to use the same tricks as they do. We may wear camouflaged clothing, or sit in a camouflaged blind, keeping as quiet and still as a moth on a tree trunk.

Soldiers also use camouflage. Like bird-watchers, they disguise themselves and their equipment. In the Arctic they imitate polar bears or ermines and wear white clothes to match the snow. In other environments they wear different colors – browns and yellows in the desert, various greens and black in the forests. In both cases their camouflaged uniform uses disruptive coloration to make the soldiers more difficult to see. They may also try to hide the shadows around tanks or guns by using covers made of netting.

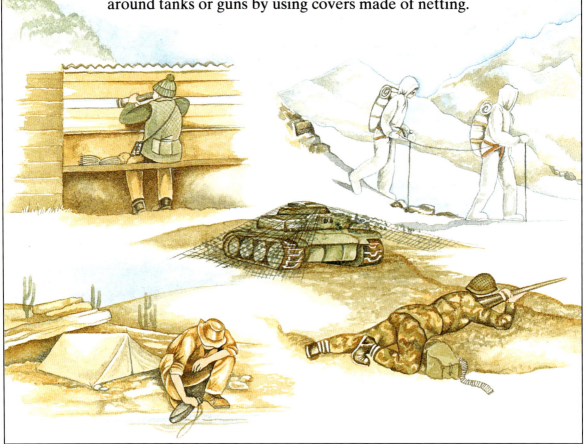

GLOSSARY

ABDOMEN The hind part of an insect, or the belly of a vertebrate animal.

AMPHIBIAN A vertebrate animal. Generally the adults are land-living and breathe air, using small lungs. Some also breathe through their skin. The young are water-living tadpoles which breathe through gills.

ANTENNA (plural: ANTENNAE) The feelers on the heads of insects or other animals without backbones. They are involved in the sense of smell as well as touch.

BACTERIUM (plural: BACTERIA) Some of the smallest and simplest of all living things. The single cell that forms the body of a bacterium does not have a nucleus or other specialized parts, like the cell of, for example, an amoeba. Bacteria are found in almost all habitats, where they are of great importance, for they are recyclers, breaking down dead plants and animals and returning them to the environment.

CAMOUFLAGE An object or animal that is difficult to see because its color and/or shape matches the background.

CARNIVORE A flesh-eating animal.

CELLS The microscopically tiny units of which plants and animals are made.

COLD-BLOODED An animal whose body temperature is dependent on the warmth of its surroundings. On a cold day it will have a low temperature; on a warm day its temperature is far higher than that of a mammal. Because of this, it does not have a steady output of energy, but it needs very little food compared to warm-blooded creatures such as mammals and birds.

ECHO A sound bounced off an obstruction. The process of locating objects by means of sound waves is called **ECHOLOCATION**.

FERTILIZE To bring together male and female sex cells so that a new generation is formed.

FOOD CHAIN The complex of plants and animals on which a creature depends for its food. It includes not only those that the creature eats, but those that affect the growth or survival of its food organisms.

GLANDS Small parts of an animal's body that release chemical messengers into the bloodstream. The effect of these is to alter the animal's behavior in some way.

INNER EAR The inmost part of an animal's hearing apparatus that lies inside its head.

IRIS The colored part of a vertebrate's eye that surrounds the pupil.

LARVA (plural: LARVAE) The young of some animals. Larvae are able to fend for themselves, but they look different and live and feed differently from their parents. When full -grown, they change fairly rapidly to the adult form.

MAMMAL A warm-blooded, air-breathing animal with a backbone, fed in the early stages of its life on milk produced by its mother.

NECTAR A sugary substance produced by flowers to attract pollinators.

NERVE ENDING The end of a nerve. It usually lies in or just below the skin. When stimulated it informs the brain of a change in, for example, pressure, light, and sound.

PARALYZE To affect the nervous system of an animal in some way so that it cannot move.

PIGMENT A chemical that gives color to animals, plants, or other objects.

PLANKTON Animals and plants that float in the currents in lakes or oceans. Most are small, many are the young stages of larger animals.

POLLINATE To transfer pollen from one flower to another.

PREDATOR A hunter.

PREY The animals caught and killed by predators.

PUPIL The dark area in the middle of the eye of a vertebrate, through which light passes. The pupil is often rounded in shape but may be oblong, or even zig-zag shaped.

REPTILE An air-breathing vertebrate animal, with a hard, dry skin, often armored with scales or bone. Reptiles are cold-blooded and their young usually hatch from eggs, though a few kinds give birth to live young.

RETINA The light-sensitive area at the back of an eye.

RODENT A mammal with two strong front teeth in its upper and lower jaws, which it uses for gnawing. Squirrels, mice, and porcupines are all rodents.

SENSES The special powers that animals have to discover things happening about them. We generally talk of five senses, sight, touch, hearing, smell, and taste. Some animals have extra senses besides these. Many animals have senses far better developed than ours. For instance, a dog has a sharper sense of smell, a bird has better eyesight.

SWIMBLADDER A gas-filled bladder inside a fish. This acts like a built-in lifebuoy, so that the fish is weightless in water. As a result, all of the energy used in the fish's movement goes to push it forward, rather than keep it up in the water.

TENTACLES Long, slender, arm-like parts of some kinds of animals. They are used for feeling and holding things, and sometimes for moving.

VENOM Poison produced by animals. It is usually injected into prey creatures to kill them, but it may be used defensively.

VIBRATE To move rapidly and continuously to and fro.

INDEX

Illustrations are indicated in **bold**

A TEMPLAR BOOK

Devised and produced by The Templar Company plc
Pippbrook Mill, London Road, Dorking, Surrey RH4 1JE
Copyright © 1991 by The Templar Company plc

PHOTOGRAPHIC CREDITS